MONEY IS NOT ALL
THAT MATTERS

*Strategies For Attracting & Retaining
Technical Professionals*

Money Isn't All That Matters:
Strategies For Attracting & Retaining Technical Professionals

© Copyright 2000 by Blessing/White, Inc.

For information, address:

Blessing/White, Inc.
23 Orchard Road, Suite #2
Skillman, New Jersey 08558-2609
USA
Phone: (908) 904-1000
Fax: (908) 904-1774
E-mail: info@bwinc.com
www.BlessingWhite.com

Blessing/White, Inc.
55 King Street
Maidenhead
Berkshire SL6 1DU
United Kingdom
Phone + 44 1628 771999
Fax   + 44 1628 645000
E-mail: info@uk.bwinc.com
www.BlessingWhite.com

First Edition

ISBN Number: 0-9678321-0-1
Library of Congress Card Number: 99-069953

Money Isn't All That Matters

Strategies For Attracting & Retaining Technical Professionals

By
Dan Treadwell & Peter Alexander

DEDICATION

This book is dedicated to all our clients
who have trusted Blessing/White to help
motivate and develop their employees
over the last 25 years.

Table of Contents

FOREWORD

Many organizations are either having trouble attracting and retaining technical talent or worried that they will lose some of their existing staff as the economy continues to flourish. And as salaries continue to escalate, money tends to be the focal point in the majority of attraction and retention efforts.

Unfortunately, attracting employees using dollars alone is the market equivalent of a bargain store attracting customers using low prices alone. The customer is only loyal until your competitor offers a lower price. The same goes for employees. They will only be loyal to you until someone else offers them more money.

But lower salaries should not be an excuse for losing technical talent. Some organizations have been paying less for years yet you couldn't steal their people for twice the pay. Why? Because they offer their technical employees some intangible benefits that outweigh monetary rewards. These benefits fall under a retention strategy that includes key motivators such as the opportunity to develop their skills and work on the latest technology, and the autonomy to make meaningful contributions to the organization (and the recognition that comes with it).

As the competition for technical professionals continues to be among an organization's greatest challenges, an effective attraction and retention strategy will be a critical success factor. Just as it is much less costly to retain customers than find new ones, the same is true for technical professionals.

If you want to keep your technical talent, put systems in place that will allow these hard-working individuals to contribute all that they are willing and capable of. To maximize a technical professional's energy, enthusiasm and creativity, make your workplace somewhere they want to be.

Christopher Rice
President & CEO
Blessing/White, Inc.

WHY THIS BOOK?

"The most important part of turnover is the loss of intellectual capital. We lose time, we lose productivity, and we lose efficiency. How do you put a number on that? I don't know, but I know it's not good."[1]
- Ted Kastelic, Intel Corporation

"Attracting and retaining top talent is one of our most important competitive advantages in the new world economy."[2]
- Barbara Beck, Cisco Systems, Inc.

A *Fortune* Magazine article titled "What Makes A Company Great" included a survey of the world's most admired companies. In the survey, this question was asked: "What is your single best predicator of overall excellence?" The answer: "The company's ability to attract, motivate and retain talented people."[3]

This is one of many articles about the talent shortage that have recently appeared in national and local publi-

---

1   Ted Kastelic, Management and Organizational Development Manager, Intel Corp. (Santa Clara, California), January 2000.
2   Barbara Beck, Senior Vice President, Human Resources, Cisco Systems, Inc. (San Jose, California), January 2000.
3   (ISSN 0015-8259) Vol. 138 No. 8 page 206+ Features/Global

cations. Everywhere you turn, you hear about how the shortage of technical professionals threatens a company's ability to compete, maintain growth, seize opportunities and secure venture capital. Companies who want to win have intense time pressure; speed to market in some instances can mean the entire ballgame. Software engineers are pushed to publish products by unreasonable deadlines, and have grown accustomed to pulling all-nighters near launch time.

> *"The competition for talent is the tightest*
> *I've ever seen."[4]*
> - John Nobrega, Raydiate, Inc.

These situations are directly affecting the bottom line. According to the Meta Group, the labor shortage is expected to cost US industry $500 billion in business revenue, $10 billion in business income, and $15 billion in increased compensation costs.[5]

> *"The foundation and growth of our business is based*
> *on how we expand on the Internet, and we need the*
> *talent to develop the Internet infrastructure."[6]*
> - Steve Schloss, Time, Inc.

The scary thing is, the dollar figures quoted by Meta Group might actually be understated! How is that possi-

---

4   John Nobrega, President, Raydiate, Inc. (Pleasanton, California), January 2000.
5   Meta Group Inc. 1999 *Worldwide Benchmark Report*, www.meta-group.com.
6   Steve Schloss, Human Resources Director, Time, Inc. (New York, New York), December 1999.

ble? Well, take for example the tangible recruiting costs for a technical professional: advertising expenses, search-firm fees, time lost due to interviews, recruiting materials such as videos, brochures, CD-ROMs, Web-page design, candidate travel, meal and lodging expenses.

Once hired, there's the cost of training and training facilities, the participant's time spent away from his job, and the costs of management's time required to coach and mentor a newcomer. In addition, there is the loss of marketplace reputation and credibility with customers because of high turnover. Finally, there are the costs of newcomer mistakes due to inexperience.

Organizations can therefore spend an average of $30-50,000 per new hire once all the tangible costs are added up. For every 10 employees who don't have to be replaced, therefore, an organization can save up to $500,000. In large companies hiring several thousand new employees each year, the reduction of turnover by only a few percentage points can mean millions of dollars in savings.

"(Technical professionals) can work for anyone they want, where they want, for as much as they want."[7]
- William Glover, Revenue Systems

And this does not even include certain technical talent considered irreplaceable by the organization. For example, one big semiconductor company was concerned

---

7   William Glover, CEO, Revenue Systems (Atlanta, Georgia), January 2000.

about losing its most gifted engineer. They determined that this one person was worth $29 million to the company. When the CEO found this out, he ordered a check written to the engineer for $1 million! As incredible as that seems, with the writing of that check, the CEO saved $28 million for his organization. And a large software company is rumored to have waved a $1 million signing bonus at a senior software developer.

Now, obviously money was the biggest factor in these situations. Yet from our research, we have found that money isn't all that matters to the technical population. It's important, of course, but there are a whole host of other reasons (both tangible and intangible) that influence a technical professional's decision to work for an organization. Examples include their first impression of the organization, the person's relationship with his or her manager and the opportunity to work with the latest technology.

As you read the information included in this handbook, you may think "this is true for all talented people, not just technical people." You are correct in this assumption. With the integration of technology in nearly every facet of business, there is a blurring of distinctions between categories of knowledge workers. We all are becoming free agents in some way, with the desire to choose how we do our work and whom we work for.

It's just that technical professionals are most in danger of being "cherry-picked" by another organization. Because they know what the market can offer, technical professionals have come to expect their employers to offer programs specifically focused and dedicated to their needs. So while technically focused, many of the strategies we discuss will also work for other employees within your organization with little or no modifications.

The attraction and retention situation is likely to get worse before it gets better. However, the strategies listed in this book will give you a fighting chance to find and keep technical professionals and directly improve your organization's bottom line.

"The biggest issue I face is staffing. It's impossible to find the right people, because everybody is looking for the same people. The demand is increasing and the supply is decreasing. Just basically, companies are stealing people from each other."[8]
- Anonymous Participant, Blessing/White Focus Group

---

8 Anonymous Participant, Blessing/White Focus Group of Directors With 150+ Technical Staff, September 1997.

WHAT YOU NEED TO KNOW ABOUT THEM

"To the optimist, the glass is half full. To the pessimist, the glass is half empty. To the technical professional, the glass is twice as big as it needs to be."
- Anonymous

The population of workers known as technical professionals has come into prominence over the past decade and represents a significant portion of today's workforce. Ten years ago, technical professionals worked almost exclusively in technology-oriented organizations. Now, virtually every organization has staffs of technical professionals who are integral to its success.

What Is a Technical Professional?

What characterizes a technical professional? The extreme example is a programmer working alone in a remote, controlled environment, focused exclusively on one specific goal. In reality, this stereotypical definition is far too narrow.

In fact, this talented population has been trained in one of dozens of linear, problem-solving disciplines such as software design, hardware design, engineering, science, aerospace, financial disciplines, medical and health-care related fields. These disciplines typically have a strong procedural and process orientation. A technical professional is any individual working for an organization either as an employee or contractor who possesses technical skills critical to the success of the organization. In other words, these are individuals whose absences can immediately affect the way your organization remains competitive.

Technical professionals are knowledge workers who bring unique values and expectations to the workplace. Yet this diverse body of people shares some common traits and common needs.

What Are Technical Professionals Looking For?

What do technical professionals look for in a job? If an organization is seeking to attract qualified technical professionals, what do they need to know to attract them? What will keep them engaged and energized? What will build loyalty between them and the organization? When an attractive opportunity presents itself, what will entice them to stay?

In our experience, several factors are critical to attract and ultimately retain quality technical talent.

Challenge

More than any other population, technical professionals keenly desire challenge. They want the challenge of interacting with the newest tools, processes, methodologies and advances. In fact, they abhor repetitive tasks and outmoded technologies. Keep them active with challenging projects and interacting with others who can offer new perspectives, and their job satisfaction will remain high.

Development

"With information technology moving ahead at the rate of a new product or upgrade roughly every 18 months, it is imperative that people keep their job skills up to date."[9]
- Donna Senko, Microsoft Corporation

One reality that must be faced with technical professionals is that their discipline is their passion. They may appreciate the organization and its work, but for virtually all technical professionals this appreciation will place a distant second to advancing their own knowledge. They have a strong need to develop and refine their skill set. In short, if they can't learn, they'll leave. Count on it.

---

9   Donna Senko, Director of Certification and Skills Assessment, Microsoft Corp. (Redmond, Washington), January 2000.

Contribution

In order for technical professionals to remain interested, they must feel they are making measurable, meaningful contributions. These highly trained, skilled employees lose interest unless they see that the work they are performing is contributing to the greater good of the project. Maximum job satisfaction is achieved when it is clear that the work being performed is vital to the success of the objective – whether for a client, the company or a cause.

Relationship

"People leave managers, not companies. So much money has been thrown at the challenge of keeping good people – in the form of better pay, better perks, and better training – when, in the end, turnover is mostly a manager issue. If you have a turnover problem, look first to your managers."[10]
- from *First, Break All the Rules* by
Marcus Buckingham & Curt Coffman

Even though many technical professionals may appear to be aloof or off in their own world, don't be fooled. They crave the stimulation they receive when interacting with other talented people in their peer group. The opportunity to share ideas, discuss different approaches and brainstorm solutions is very attractive to them. The

---

10 *First, Break All The Rules* by Marcus Buckingham and Curt Coffman, Simon & Schuster, 1999, p. 33.

same applies to their relationships with their managers. Official employment statistics clearly state that the number one reason employees leave their jobs is that they don't get along with their managers. Nowhere is this truer than with technical professionals. By the same token, when they do receive support and encouragement from their managers, they thrive and are reluctant to make a change that might jeopardize that dynamic.

Technical Professionals Have Unique Needs

A recent mailing targeted to human resource professionals heralded, "One of these 7 things will motivate any employee in the company." It stated, "99% of employees are motivated by one of the following 7 needs:

1. The need for achievement

2. The need for power

3. The need for affiliation

4. The need for autonomy

5. The need for esteem

6. The need for safety and security

7. The need for equity"[11]

---

11 Adapted from *The Manager's Desk Reference* by Cynthia Berryman-Fink & Charles B. Fink, the American Management Association.

Makes sense, doesn't it? We all know people who are motivated by any one of these needs. What makes working with technical professionals different and challenging is that they have *several* needs that must be addressed at the same time to maintain a high level of job satisfaction.

Blessing/White made this discovery in the late 1980s when we introduced a skills-based leadership program developed in partnership with several existing clients. The program was highly successful, and the clients who initially implemented the process were pleased. It was only when the program was introduced at a new client's site that we experienced our first failure. It wasn't that the audience didn't like the program; they HATED it! Evaluations for the pilot session came back with comments like, "this is not about us," "not appropriate for our environment," and "what, are they kidding?"

Within six months, two similar experiences occurred and motivated us to examine what these three failures had in common. The commonality? They all occurred in technology-based organizations or technology-based departments within an organization.

As a result, a three-year study of 19 technology-oriented organizations was launched. The objective of the research was to identify and analyze strategies and behavior patterns that made technical leaders successful. Each of the 19 organizations was asked to identify the technical leaders who were successful from the organization's perspective and highly regarded by the professionals whom they led. The research population included professionals in computer hardware and software, engineers and scientists from various industries.

These leaders (and the professionals whom they led) were interviewed and observed. The research identified in no uncertain terms why the prior leadership training failed, why it was "not about them."

Six distinct needs were identified and validated by the research findings. Effective technical leaders must be able to understand and respond to these needs of technical professionals:

Autonomy

Technical professionals are achievement-oriented individuals who derive motivation from the work itself. A high level of autonomy over the conditions, pace, and content of work is important to them and they are increasingly sensitive to the quality of the work environment and culture. The need for autonomy usually means a higher need for participation in goal setting and decision-making. At the same time, they have a strong need for self-management. In fact, most technical professionals prefer a leadership style that gives them as much independence as possible.

"Technical professionals are bright, independent and achievement-oriented individuals. It is critical that we respond to their needs as professionals in order to insure that we retain and develop them for future growth." [12]
- Lawrence J. Krema, NEC Technologies, Inc.

---

12 Lawrence J. Krema, Vice President, Human Resources, NEC Technologies, Inc. (Itasca, Illinois), January 2000.

Achievement

Technical professionals are driven by a need to accomplish goals requiring a high degree of skill and effort. They want to accomplish something of major significance. Their sense of achievement is enhanced when they understand how their work relates to the organization's goals. Work that is perceived as exciting for the technical professional and important to the organization generates commitment, as does the opportunity to put one's skills and knowledge to the test.

Keeping Current/Avoiding Burnout

"We help our technical people deal with the high rate of change associated with our products and customer expectations by providing ways for them to learn new skills and use them on the job."[13]
- Tom Kristoph, PRI Automation

Technical professionals fear obsolescence. Underutilized skills often yield apathy, burnout, and/or alienation resulting in interpersonal problems, demotivation, and declining performance. Burnout exists when the professional has lost the sense of purpose from the work, is emotionally exhausted, and feels trapped and powerless to influence change.

---

13 Tom Kristoph, Director of Development & Training, PRI Automation (Billerica, Massachussetts), December 1999.

Professional Identification

Technical professionals tend to identify first with their professions and second with their organization. "Professional first and employee second" characterizes many specialized technical people. As a result, the pursuit of professional goals can conflict with the attainment of departmental and organizational goals.

Participation In Mission and Goals

Technical professionals are more resistant to committing to mandated organizational goals than are most occupational groups. Since acknowledgment of their input is so vital to these individuals, participation in the goal-setting process is key to their motivation and job satisfaction. Alignment of their goals with those of the organization is fundamental to establishing and sustaining motivation. However, once committed, technical professionals often set high performance standards and can experience anxiety over attaining them. They can also develop so strong an attachment to goals and standards that change becomes upsetting and demotivating.

Collegial Support and Sharing

The potential for competition is high among these bright, ambitious people with strong egos. At the same time, collegial support is important to technical professionals. They value interacting with others who posses knowledge and experience different from their own. They favor a collegial relationship with their leaders as well. Open discussion of goals and a collaborative approach is highly motivating to this population.

But aren't these needs true for everyone at one time or another? What makes these needs so relevant to technical professionals is the degree to which they coexist at the same time. Technical professionals are not motivated by just one of these needs, they're motivated by up to all six needs at the same time. And these needs aren't always in harmony with each other. Some paradoxes exist for technical leaders. For example, what's the best management approach for a professional who craves autonomy and collegial support and sharing?

Special Requirements Of Technical Leaders

"We want our technical managers to have a sense of ownership in keeping their employees motivated and engaged."[14]
- Ben Putterman, Oracle Corporation

Given the strong needs of technical professionals, managing them requires special knowledge, strategies and tactics. This would present a substantial challenge even to leaders with high levels of interpersonal skill and

---

14 Ben Putterman, Director of Worldwide Management Development, Oracle Corporation (Redwood Shores, California), January 2000.

aptitude. Yet technical leaders rarely possess a combination of technical expertise and leadership skills. This is not surprising given the complexity of most technical disciplines where education leaves little, if any, opportunity to focus on interpersonal dynamics. Most often, technical leaders are promoted to management based primarily on their technical competence and most often work for someone whose focus is on technical results and not interpersonal relationships. As a result, technical leaders often lack the benefit of adequate role models.

Typically, the majority of today's technical leaders are technical professionals whose abilities, personalities, and interests are more "process" than "people" oriented. While superior technical skills can influence early success as a leader, interpersonal effectiveness plays a crucial role in determining a technical leader's long-term achievement. The focus for technical leaders is typically not on managing people, but on managing the technical aspects of projects, that with which they are most familiar. However, the increasingly fast-paced, competitive world of technology requires well-rounded leaders who are responsive to the needs of technical professionals as well as to the attainment of the organization's strategic objectives. This complex leadership requirement combined with the population's unique needs requires a unique approach and sensitivity. Happily, effective leadership methods can be learned. And technical leaders are some of the quickest learners there are.

What They Need to Know About You

Paternalistic organizations are essentially a thing of the past. The notion of being a "lifer" in an organization (retiring from the same organization you joined out of school) is virtually unheard of. The "safety nets" earlier generations have relied on (unions, pensions, contracts, and tenure) are rapidly disappearing.

The tables have turned. In an environment where the majority of us are "employees at will," many organizations (especially those that are technology-based) are recognizing the need to convince key employees to stick around by promoting the benefits of working for them. These organizations realize the importance of helping each employee clearly understand what's in it for her/him.

The economy is good. Technology is booming and technical professionals are in demand. Here's the grim reality: we need them far more than they need us. In the past, we may have assumed the benefits of working for our organizations were obvious. Yet "employees at will" are essentially free agents – and their eyes are open. If they can no longer find a paternalistic employer who will make a long-term commitment to them, they seek employment opportunities where they can further their career, sharpen their skill sets and work on projects they find interesting. So, they scan the classifieds on a regular basis. They go on-line to see what's "out there." They're smarter than they're given credit for and they're often dangerously close to leaving – this is particularly true of technical professionals.

So, how do we attract and retain this elusive technical professional?

What Do Technical Professionals Look For In a Company?

What are some of the fundamental criteria talented technical professionals use to determine if our organizations are worth investigating?

Opportunity

Opportunity for what? For growth, for advancement, for recognition and for achievement. If they're not convinced that we will offer them opportunities to grow and prosper, they're not interested. They don't care about where we've been, they're only interested in where we're going. Will our organizations give them the opportunities they're looking for to pursue their technical discipline?

Challenge

Once they're on board, how will our organizations keep them challenged? Are our work and approaches innovative? Do we listen to the feedback from our professionals? Do we act on it? Do we respect our professionals and inspire them to do their best work?

Working With Leading-edge Technology

Are our organizations doing cutting-edge work? Do we have state-of-the-art equipment? Are we defining procedures and protocols? Are we attracting the best minds? If not, why should they work for us?

What Do Technical Professionals Need To Know About Your Company?

"The young recruits we wanted to bring in didn't want to join a company. They wanted to join a vision and have an identity in the firm. A personal feeling of making a difference."[15]
Rick Throckmorton, Booz, Allen & Hamilton

So, let's say your company has made the first cut – you're a likely candidate in the technical professional's view. They've investigated you and see opportunity, challenge and leading-edge activity. Now what do they need to know? What will convince them that your organization is one they want to join?

---

15 Rick Throckmorton, former Vice President of Booz, Allen & Hamilton (McLean, Virginia). *InformationWeek Roundtable,* April 20, 1998.

Be prepared to build their interest by giving them information about:

Mission/Direction

Where is your organization going in the next year? In the next three years? Why have you chosen that direction? What gives you confidence that you are heading in the right direction? How do you stack up against your competition?

Values

What does the organization stand for? What are the most important criteria used when making difficult business decisions? What motivates management? Furthering technology? Shareholders return? Pursuit of organizational goals? Market share?

Viability

Is this organization a going concern? Do you have the funding to support your growth goals? For the short term? For the long term?

Policies/Benefits

What's it like to work for your organization? What kind of flexibility do employees have? (Remember that need for autonomy!) Is training in my functional discipline available? Will you fund advanced degrees? Do you have a sabbatical program? How will you support me in keeping in touch with the state-of-the-art?

Salary/Bonus

How are my results rewarded? Is your salary structure low or high in comparison with others? How is my discretionary effort recognized and rewarded? Is any portion of my compensation (salary and/or benefits) tied to the performance of others? Other departments? Company results?

Work/Life Balance

What do you do to help me maintain a healthy balance between my work and personal life? Are day care facilities available? Will you be flexible if I have family-related needs? Do you have a health club on site or fund a near-by facility? Are there any wellness benefits available?

What Do They Want To Know About Their Potential Leader?

Leaders and managers represent the organization on a day-in, day-out basis. The relationship with the leader is arguably the most important association any employee has in the workplace. In order to feel good about this primary relationship, what information do technical professionals need from their managers?

- Manager's style

- Manager's background

- Performance goals – what you need from them

- Your ideas on how to best communicate with each other

- How you can help them

- How receptive you are to their opinions

What Will Motivate Them To Stay?

In some cases nothing will motivate a technical professional to stay. Sometimes another offer is just too good or unique. Or – and we always hope this doesn't happen – so much damage has been done to the relationship that nothing will compel them to stay. Yet there are MANY things enlightened leaders and organizations can do to build healthy relationships and ensure that technical professionals are nurtured in the right ways and encouraged to stay.

Job Satisfaction

When technical professionals are highly satisfied with their jobs, they are much more inclined to stay with the organization. Job satisfaction can be defined in a million different ways. For one professional it may focus on compensation, for another work/home balance, for another challenge, and so on. One thing is for sure, for leaders to help a professional manage his or her job satisfaction, they've got to get to know them as individuals. Talk to them and actively seek to understand what they will find satisfying. Simply increasing the frequency and quality of communication with a technical professional can significantly increase job satisfaction.

Partnership/Collegiality

It is harder for anyone to leave an environment where they feel supported and part of a team than it is to leave when they don't feel any connection to the manager and co-workers. Technical populations in particular value collegiality and the esprit de corps that comes when people are working well together and stimulating each other to do their best work and achieve the best results.

Positive Working Environment

Leaders have a unique opportunity to create working environments that inspire professionals to do their best work, environments that build loyalty and trust. Often key managers leave technical organizations and their staffs follow them. Why? The positive working environment they built took on an energy of its own. An energy so attractive that they became Pied Pipers.

Room To Contribute

When technical professionals feel they are making a meaningful contribution to an organization, they are more likely to stay. They feel needed. They feel they are necessary and an integral part of making success happen for the organization. Of course, significant contributions need to be acknowledged with fair (even generous) tangible rewards, yet the intrinsic reward of respect and feeling valued has proven far more compelling than monetary or material rewards and perquisites.

Valued Opinion

Everyone wants to be a valued contributor to the organization. When technical professionals' points of view are heard and recognized, that in itself serves as a reward, even if the point of view triggers disagreement or debate. In fact, debating a professional's point of view respectfully and offering different perspectives and ideas is stimulating and valued by this "critical" and "analytical" population. True, they may walk away shaking their heads, but more often than not, they will come back another day to debate again if they feel heard and if the debate has a positive intent.

ATTRACTION STRATEGIES

E ffective attraction of technical professionals has become a critical element in the financial success of an organization. If you make a mistake in hiring, and you recognize and rectify the mistake within six months, the cost of replacing the employee is 2.5 times the person's annual salary. And that does not include the emotional costs.[16]

As the title of this book indicates, when it comes to accepting an offer, money will not be the only thing that is considered by the potential recruit. Therefore, we have organized the attraction suggestions of this chapter into sections based on the primary and secondary research we have conducted over the last several years. These sections are as follows:

- Personal flexibility and convenience – options that fit their lifestyle.

- Professional challenge – always learning and growing in their area of interest.

- Competitive compensation – all things being equal, the technical professional wants to be fairly compensated.

---

16 *Inc.* Magazine, August 1998, "No Room for Compromise" – Pierre Mornell, www.hiringsmart.com.

- Creative venues – places and tactics you may not have thought of.

- Virtual strategies – attracting technical professionals outside your own backyard.

- Referral programs – tapping into your employee's talent network.

- Interviewing tips – dos and don'ts for this unique population.

It is our hope that by reading these strategies you will be able to generate an idea or two that will help you attract technical people to your organization. And, as mentioned in the "Why This Book" chapter, many of these strategies will also work for talented employees outside of your organization's technical staff with little or no modifications.

Personal Flexibility and Convenience

1 Offer on-site day care (for those with family responsibilities), dry cleaning, concierge, ATM or banking services, haircuts, car washes/oil changes, etc. Technical professionals have a lot of stress, especially when dealing with unrealistic product release deadlines. By providing a way for them to easily take care of their errands and/or stay more connected with their children, you will have a competitive advantage in attracting their talent. Tout this information in your annual reports, and use the annual report as a direct-mail recruiting piece.

2 Stagger HR staff hours to cover all time zones. You have to be available to speak with technical professionals at hours that are convenient for them first and foremost. For example, if you are an East Coast company recruiting from the West Coast, there is a 3-hour time difference that must be accounted for. Try to have your HR staff work different hours to accommodate the different time zones, and make sure somebody is available to make a quick hiring decision if an outstanding candidate is ready to come aboard.

3 Savvy IT organizations will proactively pursue technical consultants who are tired of being on the road, have new personal commitments (recent marriages, elderly parents, etc.) and are seeking greater balance in their lives. This target audience is especially attractive, because it enables your IT group to take advantage of the substantial training invested in these consultants during the early years of their careers.

4 Technical professionals love flexibility in their schedule. They may come into the office late on occasion, but they also tend to work late into the evening either in the office or remotely from home. Give them the opportunity to telecommute when they want, and provide them with the equipment to be productive from home. You will find that most programmers work best when they have limited distractions from people in the office. Also, if your technical staff is compensated on an hourly basis, consider giving them creative scheduling options such as an 80-hour workweek spread over nine days instead of ten (every other Friday off). A three-day weekend twice per month can be a compelling incentive.

5 If the position you are offering requires the candidate to relocate, consider offering recruiting assistance to his or her partner to find a new job as well. You will eliminate one of the decision barriers for the candidate, and will also build a positive first impression with that person's significant other.

Professional Challenge

6 If you are a small business, be sure to promote an open management style, little bureaucracy, entrepreneurship and adventure. These are advantages you have over a larger organization, which can be full of red tape and politics (a major turnoff to most technical professionals).

7 When you post for open positions, write ads that, rather than describing a job opening, describe the type of person who would be successful and happy in the position. Talk about the technology and projects the person would be working on, what he or she will achieve, and how management and the organization will recognize him/her. Technical professionals want to make sure they are contributing to the organization, so make sure your job descriptions cater to this need.

8 Schedule and promote an open house event at your headquarters and other locations where there is a significant technical operation. Demonstrate your technical capabilities and discuss your company's vision. If done right, you may be able to hire talented people from the event because the candidates will experience your technology first-hand and be able to see inside your company's culture in a non-intimidating manner (as opposed to a formal interview).

9 Allow potential recruits to take a one-day "test-drive" working for your organization to see if the job is challenging enough to keep them motivated. This will give both the candidate and your organization a chance to decide if the relationship would work long-term. For example, Support Technologies, Inc. (Atlanta, Georgia) sponsors an event called "Support Technologies Experience," where they have a one-day audition for serious candidates. Candidates spend a day with the whole company, working with people as they do their jobs. The two main things the company is looking for are: 1) is he/she willing to speak out and challenge the status quo, and 2) would she/he fit in? The company is most interested in people with opinions, criticisms and concrete solutions to problems.

10 Encourage technical employees to do community service by providing paid time off in exchange for their time helping their community (for example, one hour of community service might earn one hour of extra vacation time). Not only will the employee feel personal satisfaction, but also your organization will receive positive word-of-mouth within the technical community when your technical professionals network with their peers. For example, The Timberland Company (Stratham, New Hampshire) has a program titled "Path of Service™" that offers up to 40 hours of paid time per year to volunteer in their communities.

Competitive Compensation

"If you pay peanuts, you get monkeys."
- Anonymous

11 The two-word title of this section nearly says it all. You have to be competitive with your compensation offer, but what does competitive really mean? It's been communicated to mean "within the same ballpark," but that still doesn't answer the question. In speaking with various technical professionals, competitive means "within 10 percent" of market value. That means that you stand a reasonable chance of hiring that sought-after technical professional if you offer a challenging, flexible, convenient and fun work environment and the compensation package does not dip below 90 percent of what your competition is offering. This means you have to keep constant tabs on what the market is paying for people with the skills you covet. Throw out the old HR salary scales, because they are outdated and won't keep up with today's salary expectations. If your organization must remain within specific salary levels, consider hiring bonuses to make up the difference and promise quarterly bonuses for meeting and/or exceeding objectives of the position. If you have plans to go public, you can also include stock options as an incentive. One way or another, you must be competitive with your offer or risk losing candidates to other organizations. [Note: in the Retention Strategies chapter, we will share a few creative compensation ideas for you to consider.]

12 Try to think "out of the box" when it comes to compensation. For example, Revenue Systems in Atlanta, Georgia needed five new PowerBuilder programmers and/or Oracle database administrators a month and were only getting three, at a cost of about $16,000 per head. Their recruiting program was not working. Rather than continue down that unsuccessful path, Revenue Systems innovated the idea of cutting out the headhunters and using that same money to provide each employee with their own leased BMW in the hopes that it would create for them a successful "word of mouth" recruiting and retention program. This strategy worked beyond their wildest dreams. By the time Revenue Systems delivered the first car, they had been able to hire 21 new people without a fee, saving over $300,000. And the whole program only costs $35,000 per month versus paying headhunters $50,000 for a no-go program! Revenue Systems now has about 65 people, with 55 cars delivered and another seven or so on order. They lease the cars and pay the property taxes, insurance and registration fees. There are no conditions or limitations on the use of the cars, and employees get them on day one (at least on order). The employees feel special, like sort of an elite group that shares a special experience, and most of them now let their spouse or significant other drive the car to their work on Fridays. This attraction strategy has helped *retain* valued employees as well.

Creative Venues

13 Use topic-specific news groups or bulletin boards on the Internet. Search for news groups that are of interest to your typical recruit and read through the postings to find the smartest contributors. Most recruiters won't read postings in news groups because those postings require a fundamental understanding of the technical topic in order to gain the respect of the users in that bulletin board community. Therefore, make sure you are familiar with the topic (you might get the help of someone in your IT staff!) before contacting any of the users.

14 Rent an airplane to fly a banner around a high-tech company that is laying off workers. Include your company name, Web site and the words "We're Hiring!" on the banner. Disillusioned employees can't help but notice your efforts. A hardware manufacturer did this to a PC maker and it was very successful. 1-800-SKYWRITE is a company that provides this service in the Silicon Valley.

15 Don't rely simply on print and electronic help wanted ads. Use networking, local colleges (academic and job placement offices), etc. If you do choose to use help wanted ads, pick your words wisely. Interpretation can mean everything to the technical professional. For example, Working Wounded (http://WorkingWounded.com), a Web site for disenchanted workers, has pointed out that:

- "Duties will vary" could mean "anyone in the office can boss you around"

- "Competitive salary" could mean "we're competitive because we pay less than our competition"

- "Must have an eye for detail" could mean "we have no quality control"

- "Problem-solving skills a must" could mean "this company is in perpetual chaos."

16 Consider certain target populations that are sometimes overlooked by large companies, such as 40+ computer specialists, mothers re-entering the workforce, retirees, people released from the military with some technical training, and four-year college students who never graduated. Retirees, in particular, show more loyalty to an employer who invests new training in them versus their younger counterparts, and stability is key when it comes to maintaining a technical department. Disabled citizens can be another untapped resource. For example, Protek Electronics Inc. (Sarasota, Florida) hires handicapped recruits from Easter Seals. With a little extra training, these recruits have become a tremendous and loyal asset to the company.

17 Job kiosks can act as your silent recruiting force. Adecco Employment Services (Melville, New York) uses "Job Shop," a high-tech interactive kiosk that graces malls and other heavily trafficked public spaces in 33 states. The service is free to applicants, who simply enter skills, background info and job preferences. They are automatically matched with specific job profiles, and if they qualify for an available job, then they meet with Adecco personnel to be interviewed and tested. This is a perfect example of being in the right place at the right time. People are frequently commuting or shopping, and if they have to wait for something, these kiosks provide a useful way to pass their time.

18 Traditional career fairs attract "career job searchers" who are constantly interviewing and hopping from job to job. That's why many successful organizations prefer to set up information booths at alternate, less formal venues such as home and garden shows, minor league baseball games, community events and microbrewery festivals. These venues attract talented technical people who are not actively pursuing another job, and it gives you the opportunity to meet these people in a relaxed atmosphere with limited interruption from other companies who are competing with you for their talent.

19 Take advantage of traditional media such as billboards and radio broadcasts. Don't just limit yourself to magazines, newspapers and Web sites. Also, consider using humor in your advertising messages. Organizations with a sense of humor appeal more to technical professionals. For example, Signal Corp. in Fairfax, Virginia used a couple of ads that proved effective. One was a spoof of the "Leave It To Beaver" television series that used the line: "Gee whiz Wally, Signal sure has some swell job opportunities!" Another advertisement had a picture of their president with pie all over his face, with the tagline "and you should see us on casual day!"

20 Consider sponsoring scholarships at local colleges and universities. For example, Sprint (Kansas City, Missouri) provided $202,500 in scholarships to 45 colleges and universities during 1999. Over 50% of these scholarships were in support of technical recruiting. In addition, Sprint has entered into a partnership with the United Negro College Fund to fund 10 scholarships annually at targeted historically black colleges and universities.

21 When traveling, keep an eye out in airports for people with T-shirts, hats or luggage from one of your competitors. Technical companies big and small provide their employees with logo merchandise as a small perk and way to advertise their company for free. You can strike up a conversation with this person and possibly recruit him or her to your organization.

22 Utilize internship programs. If they like your company, you can take them off the market before they are on it (but make sure you show them the fun side of work and don't work them too many hours). Interns can be college or high school level. Young workers make things more interesting for everyone because they bring a fresh perspective, and in technology they are looked on with a lot more respect than in other fields. You may also want to develop partnerships with local universities to set up campuses specifically aimed at training technical workers. For example, Sprint (Kansas City, Missouri) technical internships usually last between 10-12 weeks and mirror the actual full-time positions for which they recruit. This gives the interns and the company the opportunity to "try before you buy" and provides the interns with many educational, social and on-the-job training experiences.

Virtual Strategies

23 Dedicate at least one HR person to focus exclusively on technical professionals. Ideally, this person would work in the same location as the technical staff. As a result, this person would become an expert in the special needs of this unique population while working virtually as a member of your HR organization.

24 Tout the benefits of your location (including international and US regional offices where technical professionals could work). For example, 14

San Diego, California area high-tech, science and economic development organizations are pooling their marketing talent for a regional initiative titled "San Diego: Technology's Perfect Climate." Iowa's "Come Home To Iowa" targets former residents, and Nebraska does a "Genuine Nebraska" campaign during University home games. Nebraska's game day blitz has resulted in 3700 people visiting the Web site and 400 requesting additional information. Also consider establishing field offices in non-traditional technology locations in Southern states such as Louisiana and Arkansas. These locations are becoming national hot spots for new technology hiring because companies there have less competition for talent.

 More and more companies are doing job simulations over the Internet. These simulations help a company see if a candidate has what it takes, and they can help the candidate get a taste of what it would be like to work in that company. You can let a candidate do some of the actual work over the Web, and you can provide a video feed of the team with whom the candidate would be working. You get the best of both worlds: an opportunity to see how the candidate works with your team while minimizing the time and travel expenses of bringing candidates in to the office for first interviews or trial days. This also makes your company look "state-of-the-art" to prospective candidates.

"Our most successful ads include the headline: How Would You Like To Work From Home?"[17]
- Lynne Frazier, Intracorp

26 Don't force relocation. Technical people should be given the opportunity to work where they want, assuming the role they have can be managed remotely. Virtual management can be accomplished using technology to form virtual teams all over the world. And technical managers can use Intranet chat/posting sites and desktop video to simulate the "water cooler" experience of working in a central office.

27 Train managers to capture names and coordinates of impressive people whom they meet at conferences. Over time, you will develop a talent database. Consider sending a coupon to your database of highest-rated talent that says something like "The day you want to come to work for us, you're hired. You don't have to go through our HR bureaucracy. We will hire you instantly." Technical professionals hate bureaucracy and red tape, so by demonstrating an understanding of this issue, you will take steps toward becoming a "preferred" employer.

28 As you develop your talent database, remember to never throw a resume away. Every two years, send out an e-mail questionnaire to see what

---

17 Lynne Frazier, Vice President of Human Resources, Intracorp (Philadelphia, Pennsylvania), December 1999.

prospects have added to their resumes. Also ask for their current location, salary range, references and contact person. You never know when this person might be appropriate and available for hire. For example, a major health care organization uses software that searches the Web for resumes of technical professionals and then matches qualified candidates to appropriate openings. The company then uses a tracking system to e-mail top candidates, updating their resumes while updating the company's talent database. By moving to e-recruiting, the health care organization has cut its spending per qualified resume from $128 (the average cost of a print classified posting) to $.06, due primarily because their online efforts result in a far greater number of technical candidates than through traditional print media.

 If your company has frequent fun events, take pictures and put them on your Web site. Technical recruits from all over the world may like what they see of your culture and want to consider working for you.

 Turn your Web site into a tool for attracting technical talent. The way your Web site has been designed can make or break a technical candidate's first impression of your company. Cisco Systems, Inc. (San Jose, California) hires approximately 1,000 new employees for approximately 1,600 available positions each month, and they have invested a lot of resources in their online recruiting presence located at www.Cisco.com/jobs. Features include:

- A powerful search function that allows a visitor to look for available positions by keywords, job function and/or location.

- A system called "Profiler" that enables candidates to create a personal profile of their background in less than 15 minutes. In addition to taking the basic information such as name, address and educational background, the system will also ask position-specific, open-ended questions based on the person's previous experience.

- A featured titled "Oh no, my boss is coming!" that when clicked will pull up a page with bold information about successful employee habits, gift ideas for the boss or workmates, or a list of things to do today. This was created because Cisco determined that they receive the majority of their online resumes during working hours.

- Another program is called "Make Friends @Cisco." Within a few days of submitting their resume or profile, a Cisco employee in a similar role will volunteer to call and chat with the candidate to provide them with insights into what the work environment is like, and answer questions about the company. If the candidate is interested in pursuing opportunities at Cisco, the employee "friend" can submit their resume through the Employee Referral program.

Referral Programs

"Employee referrals and networking events are by far the most effective way for us to find technical people. We offer our employees a premium for referring someone with the IT skill sets we are looking for."[18]
- Steve Schloss, Time, Inc.

 Employee referral programs are very popular. Offer prizes such as money, Florida vacations, etc. Promote the referral program heavily within your organization. For example, Thomson Financial (Boston, Massachusetts) spent $17,000 to bring zoo animals on-site to promote their employee referral program and to give away a trip to the San Diego Zoo for the employee who referred the most people. Employee referral costs of $1,500-5,000 are a bargain compared to most headhunter fees, and the quality of the individual is likely to be better because the referring employee has a stake in that person's success. And to minimize abuse of the system, stagger payment to the referring employee to help ensure the new recruit sticks around. Other referral program examples include:

- I-Cube (Cambridge, Massachusetts) offered, in addition to $2,000, a 32-inch TV for every successful referral in 3rd quarter 1998. A VCR was tacked on if the person started in August. Annually, anyone making three refer-

---

18 Steve Schloss, Human Resources Director, Time, Inc. (New York, New York), December 1999.

rals receives a choice of either 2 mountain bikes or a year's worth of laundry and cleaning services. For 5 hires, the choice is a spa trip or an adventure vacation. And 8 successful referrals during 1998 resulted in a new Jeep Wrangler.

- Bristol Technology (Ridgefield, Connecticut) offers a $675 Cannondale mountain bike for every successful recruit.

- A computing company in Austin, Texas held a recruitment drawing: by sending in a qualified technical resume of someone they knew, entrants became eligible to win a Porsche. Out of 2,000 entries, they got 24 solid leads.

- Cisco Systems, Inc. (San Jose, California) has an employee referral program called "Amazing People" that results in 45-55% of all new hires. Through the program, employees can refer a job candidate and then track the status of their referral online.

- A major department store chain offers $100 to charities for each technical person an employee recommends. The company gets the tax write-off as well as the referral, and the referring employee gets recognized for helping his or her favorite charity.

 Develop a resource talent pool. Assess IT employees' skills and pool the information in a database that is accessible throughout your organization. Use the database as a sourcing option when vacancies occur. If nothing else, you will know whether or not someone else in the organization has the expertise you are looking for to handle some duties while you recruit for the position.

 Don't forget to ask your customers if they have anyone they can refer to you. Customers have a stake in your success, and if you hire someone they refer, you build upon the existing relationship. You can even offer them a reward. For example, Southwest Airlines (Dallas, Texas) asks their frequent flyers for names and resumes of technical candidates, offering as an incentive a chance to win a Las Vegas vacation. In addition, Southwest sometimes involves its most frequent flyers in their flight attendant group interview process.

Try promoting non-technical employees into technical positions. A person who has business skills as well as firsthand knowledge of a particular business unit can greatly benefit the IT staff's understanding of the organization. It also gives the non-technical employee an opportunity to grow his or her skills while providing a needed resource for the organization. For example:

- State Farm Insurance (Bloomington, Illinois) has partnered with the Center for Information Systems Technology (InfoTech) at Illinois State University to create an exclusive 12-14 week training program designed for non-technical employees seeking to change their career into the computer field. The program is designed to teach students how to program

State Farm's large-scale computer systems, including methodology, language, and productivity tools. Upon successful completion of the program, the employee is awarded a Certificate in Computing Technologies by Illinois State University and begins his or her new technical career with State Farm.

- A telecommunications company in Oklahoma has transformation training for payroll clerks, dispatchers, and others who want to take a 7-week course and become associate systems analysts who develop and test client/server systems.

- Blackbaud, Inc. (Charleston, South Carolina) had trouble finding qualified and experienced programmers so it created a 12-week basic training program for employees with the aptitude for programming and interest in moving into those positions.

35 Keep in touch with former employees using your database system. If they were productive employees, who's to say they can't be productive for you again? If you and the employee part on amicable terms, they can serve as a "corporate ambassador" for you in their next career. If you let them know you are looking for certain talent, they can help refer candidates your way and pre-sell your organization to the candidate. And don't rule out rehiring the former employee. These rehires, also known as "boomerangs," have the ability to hit the ground running when they return.

Interviewing Tips

36 "Instant Hiring": Do it for your top hires who are most critical and who represent the most limited pool of talent. The formal assessment and offer process has to be quick and easy. Assume that talented technical people who decide to leave their job will be on the market for one day. If you delay, you lose. In other words, if you like them and they like you, hire them on the spot!

37 Create a system that allows you to quickly complete a comparative analysis of resumes for a technical position. Make a chart listing education, experience, knowledge, skills, abilities and personality characteristics. This gives you a framework to compare candidates easily and will help you speed up the decision-making process - a critical factor when you are competing for talent with several other organizations.

38 Be flexible! Make the interview appointment convenient for them, even if it's not between 9 and 5, so they won't be nervous about being away from work. And respect each candidate's time during the interview. Candidates will use this as a criterion for whether they want to work for you or not.

39 Don't make it appear as if your employees work around the clock. If you are recruiting technical talent straight out of college, try to avoid calling students after hours or on weekends. If you do, students might get the impression that your people work late and/or on weekends, which can be an immediate turnoff.

40 Make a good first impression by keeping the office clean and orderly. If the office is a mess, paperwork everywhere, furniture tired, and/or there is a general sense of disarray, technical professionals may feel uncomfortable and deem your organization as lacking direction and resources. Remember, you never get a second chance to make a first impression.

41 Get your highest paid technical people to participate in the interview process. They know the skills needed by your organization and they can test your candidate's technical abilities. Also, coordinate questions among interviewers. Technical candidates get frustrated quickly when they are asked the same questions over and over again.

42 Ask recruits important questions like: What organizations do you belong to? What conferences do you attend? Where should we advertise to find people like you? What would your headline be on an ad to recruit technical specialists? In addition, ask candidates about other companies they're interviewing with, and what makes that particular company attractive. Query current employees about what they liked and disliked about the hiring process, and send an e-mail inquiry to people who have declined an employment offer asking why. You can use all this pertinent information to more effectively reach your technical targets in the future.

 Use videophones early in the screening process to look for visual clues and determine how well a candidate will fit in before going to the trouble of setting up face-to-face interviews. You will save yourself and your candidates a lot of wasted time, energy and resources. Bonus tip: don't forget to comb your hair!

Retention Strategies

"We recognize that IT is one of the last areas where we can have a competitive edge. Our technical people keep us in the game, and without them, we lose our competitive advantage."[19]
- Jack Yusko, Fort James Corp.

How do you know a technical professional is considering leaving your organization? Is there some sort of magic formula that will keep someone from leaving? No. But there are some behavior characteristics common in individuals who are looking to leave your organization, including:

- A noticeable attitude change
- A slowdown in communication
- Longer lunch hours
- Frequent absences
- More personal phone calls than usual
- Improved grooming and appearance
- A neater desk (employee is doing less work or is bringing home personal artifacts)

---

19 Jack Yusko, IT Human Resources Manager, Fort James Corp. (Norwalk, Connecticut), December 1999.

- Change in vacation patterns (many on the way out take vacations)
- No longer takes work home.

It's also important to realize that when a technical professional quits, you need to act fast. Solicit and listen to the person's reasons for leaving. Determine whether you have a chance to retain their services. If there is still a chance, get the president involved if you have to, because he/she has nothing more important to do than sit down with the individual. But if the technical professional has made the mental separation, it is best to let them go. Statistics indicate that once a person has separated mentally, they will likely leave the organization within a year regardless of your efforts.

"You have to have a comprehensive approach ... that's well thought out. You want to approach retention on many different levels and in many different ways. Otherwise, an individual program here or there is too fragmented." [20]
- Steve McMahon, Autodesk

The following retention strategies have been compiled over the last few years using primary and secondary research. For ease of reading, these tips and ideas have been organized into the following key areas:

---

20 Steve McMahon, Vice President of Human Resources, Autodesk (San Rafael, California). "Keep Them!" by Charlene Marmer Solomon, *Workforce* Magazine, August, 1997.

- Orientation – getting them started on the right foot.

- Communication – keeping information flowing.

- Personal & professional development – growing in and out of the job.

- Creative compensation – alternatives to payroll.

- Recognition & involvement – acknowledging contributions and inviting participation.

- Fun & engaging environment – so what's wrong with having a good time at work?

It is our hope that by reading these strategies you will be able to generate an idea or two to implement within your organization that helps you motivate and keep your best and brightest technical talent. And, as mentioned in the "Why This Book" chapter, many of these strategies will also work for talented employees outside of your organization's technical staff with little or no modifications.

Orientation

 Understand that the retention process starts when the technical professional accepts a position with your organization, not on their first working day. Develop and send a "newcomer kit" to new hires shortly after they accept an offer. Make it celebratory in tone and provide a broad picture of your firm, your key strategies and your people. The tone should reinforce the recruit's decision to join your organization, and make it clear that your organization is glad to have them. Thus, you are making an effort to retain the person right from the start.

45 Remember...you never get a second chance to make a first impression. If new technical hires are welcomed on their first day, made to feel wanted and valued, given timely and complete information on the company and their role in it, trained on essential topics, and coached and supported, their feelings about the organization are enhanced and they are more likely to stay.

46 During the technical professional's first day on the job, have his or her manager draw a "T" on a piece of paper, listing on one side what your company will do to help that person reach his or her goals. On the other side, list what the company expects in return (professionalism, productivity, punctuality, etc.). Then the manager should tell him or her that if the company is fulfilling their end of the promise, they expect the individual to remain an employee of the company. The new employee then signs the paper. This process firms up initial expectations and gets the commitment from the new employee right away.

47 Set the tone, right at the beginning, to establish a partnership between the technical employee and his or her manager. The manager should meet with the new employee on his/her first day to review the reason(s) for the job, the skills essential for success, and what should be accomplished in the first few months of employment. The manager should also talk about his or her own background and management style, and find out what the new person needs from the manager to be coached effectively. Finally, discuss areas of growth the technical employee is interested in pursuing.

 Use creativity for your general orientation meetings. If they are boring and dominated by lectures, the technical professional may feel like your organization lacks the creativity to develop new products and succeed in the future. This also shows a new recruit that having fun in the workplace is encouraged. For example:

- Hewlett-Packard (Palo Alto, California) provides Play-Doh and toy dinosaurs on a table at some meetings as a way of helping new workers visualize company goals and objectives.

- Sun Microsystems (Mountain View, California) holds an egg-drop contest, asking teams to build a contraption that will prevent an egg from breaking from six feet.

- Disneyland Resort (Anaheim, California) includes one exercise to name all of the Seven Dwarfs.[21]

 To determine the success of your orientation process for technical professionals, ask newcomers and management to evaluate the assimilation

---

21 Used by permission from Disney Enterprises, Inc.

process. Then compare the results to objective measurements such as new-hire turnover, costs and productivity rates. You can use this information to further improve your orientation program.

Communication

50 Conduct internal focus groups to find out exactly what it takes to keep technical people happy. Ask them about perks they truly appreciate and have them categorize those perks as "need to have," "nice to have" and "don't need to have." Tabulate the scores, provide the results to the participants, and then establish and communicate a plan for implementing the "need to haves" as soon as possible. Provide quarterly progress reports to your technical staff. As a double-check, conduct detailed exit interviews to find out what the dissatisfiers are, add these to the quarterly reports, and make changes to rectify the problems where appropriate. Your technical staff will feel like your organization listens to their needs.

51 If your compensation plan is the best or near the best in the industry, make sure you communicate that effectively and frequently with your technical staff. Provide them with salary reports and studies pertinent to your industry. Make sure they know that the grass is not necessarily financially "greener" on the other side of the fence.

52 Reduce your turnover rate by giving technical employees frequent information on other career opportunities in your organization. That way, if they tire of their role, they can consider other positions within your organization rather than moving on to another employer. Remember, avoiding burnout and keeping current are key concerns of technical professionals.

53 Consider setting up your company intranet for monthly online "chats" with senior management, and allow employees to log on anonymously for no-holds barred conversation. The information that will be provided anonymously will give you a strong sense of the current state of employee satisfaction, and potentially give you some ideas for improving morale.

54 Establish a "mistake of the month" program. Set aside 30 minutes during a monthly meeting of technical workers to review mistakes and vote on those they've learned the most from. Offer prizes such as "mock" trophies or a coveted parking space for the next month. It will improve morale and open up communication within the group.

55 Because technical professionals frequently work long hours due to tight deadlines, their personal and family life may, at times, become a secondary priority. Enlightened organizations realize that a family can provide the necessary support that a technical person may need in times of stress, and therefore these organizations communicate and acknowledge the role the technical professional's family can play. For example, your company might send a gift basket home to a technical worker's family thanking them for their support during times when the employee has to work long hours, evenings or weekends. Thoughtful gestures like this will likely make the family more tolerant of the time commitment required for the job.

Personal and Professional Development

"Thinking outside of the box and creating new, meaningful challenges for our technical employees will go a long way to reinforcing their commitment to the mission, goals and strategies of the organization."[22]
- Lawrence J. Krema, NEC Technologies, Inc.

 Establish a career development program that offers opportunity for advancement into management or non-management positions as well as a clear career development plan. Include different work experiences, reading, participation in the community, learning outside of office hours, and regular practice. Determine the success of the program by measuring increased productivity and reduced costs related to turnover (assuming the turnover rate has decreased).

---

22 Lawrence J. Krema, Vice President, Human Resources, NEC Technologies, Inc. (Itasca, Illinois), January 2000.

57 Establish a dual career ladder system: one as a management ladder and the other as a technical ladder. Make room for high-level technical specialists who, although not on the management track, are compensated as well as or better than managers. That way you don't force technical professionals with little interest in management (remember, they like autonomy and limited red tape) to take on management roles just to further their career development.

58 Promote dialogue between the technical professional and his or her manager. Encourage technical professionals to "take control" of their careers by actively partnering with their managers in designing their own career paths within the organization. This will help the technical professionals clarify their importance to the organization as well as understand how their contributions relate to the success of the business (remember, they must feel they are making measurable, meaningful contributions).

59 Because of the emphasis technical people put on self-reliance and the speed with which technology changes, offering training ranks high in importance. Most technical workers realize their employability depends on keeping their skills up to date. Training means survivability to them. If you offer them the opportunity to continuously increase their skill set, technical professionals are less likely to consider leaving your organization.

"Establishing a corporate university is one of our top 5 corporate objectives for 2000."[23]
- David Owens, St. Paul Companies

 Establish an in-house corporate university. Companies like Unitel (McLean, Virginia), I-Cube (Cambridge, Massachusetts), and CoreTech Consulting Group (King of Prussia, Pennsylvania) say their in-house universities provide the following benefits:

- Improved recruitment – on-site training relevant to their position is an attractive benefit to potential employees.

- Increased revenues – through higher motivation of employees.

- Reduced turnover – providing relevant training means the technical professional does not have to search elsewhere to stay current.

- Better employee advancement – after 90 days at Unitel, newcomers are eligible to become freshmen at their university and take several classes. If they pass their freshmen requirements, they are eligible for up to an 8% raise. Additional 8% raises are available after completing sophomore, junior and senior levels as well.

- Wider talent pool – I-Cube has a 5-week program called I-Attitude that has allowed the

23 David Owens, Vice President of Corporate University, St. Paul Companies (St. Paul, Minnesota), December 1999.

company to hire workers with little experience and give them the technical training they need to serve their clients. Once completed, there are 80+ courses to help them develop and move up the ladder.

• Improved client satisfaction – One of CoreTech's CTU (CoreTech University) offerings is a 40-hour Project Management Training and Certification Program that has helped employees to deliver over 90% of projects on time and under-budget to clients.

 In addition to training, offer other ways for technical workers to keep their skills and knowledge current. One example would be to start a professional development club. Assign each of your technical employees a trade publication to read and share any interesting stories, trends or new products that they uncovered with the rest of the department. Another example is to sponsor monthly luncheons with guest speakers. Poll your technical employees to see what information would be most useful to them. Arrange for a guest speaker to talk briefly on that topic. Also establish an audiotape lending library for self-development. The more creative you are about helping a technical person keep current, the less likely it is that he or she will consider working for someone else.

"I found, especially with technical people, that the more toys you give them, the happier they are. They aren't necessarily motivated by salary. If there's new technology out there, they want the newest, biggest, fastest equipment to work with."[24]
- Anonymous Participant, Blessing/White Focus Group

 Create technology playgrounds at work (relevant to business or not). This gives technical professionals the opportunity to do interesting work and learn new skills, especially if your organization is in a non-technical industry. Or let people work on projects they select on occasion. For example, Genentech (South San Francisco, California) offers to let scientists work on projects of their own choosing a few hours per week.

 Establish a formal mentoring program for technical professionals. Allow the mentee to select the mentor based on information the mentor has provided online about personal goals and interests (rather than name, title or salary). Mentors need not be at a higher title or salary level than mentees, because technical professionals often know more than their managers in specific areas. And don't limit mentoring to corporate headquarters. Virtual mentoring can work worldwide via the Web. Not only will the technical mentee be exposed to new knowledge, the mentor has the opportunity to share his or her knowledge. This promotes a strong sense of partnership and collegiality.

---

24 Anonymous Participant, Blessing/White Focus Group of Directors With 150+ Technical Staff, September 1997.

"Our ability to offer technical professionals the opportunity to build products that utilize next generation technology has been our most successful retention strategy."[25]
- Ben Putterman, Oracle Corporation

 The best large companies have learned to mimic qualities of small companies by creating smaller, more autonomous units. These units are created within the larger organization and can tap into the financial resources of the parent company. Thus, these units can afford to give a technical professional more responsibility (a characteristic common in small companies) and a bigger budget than they would likely receive at a smaller company. This combination, coupled with the stability of a larger organization, can be very enticing and will help retain certain technical talent.

 If you are hiring for a position that maintains an outdated system, guarantee new technical recruits that within 18 months they won't be doing maintenance. Rather, they will get the opportunity to work on newer technology. This will alleviate some of the concern a technical person may have about his/her skills becoming obsolete.

 When work pressures pile up, consider establishing a pool of contingent workers as a solution for temporary and part-time staffing needs as well as a way to offer more balance to your technical

25 Ben Putterman, Director of Worldwide Management Development, Oracle Corporation (Redwood Shores, California), January 2000.

employees. Recruit contingent workers from current retirees who are interested in continuing their employment on more non-traditional terms as on-call, contract employees.

67 If your organization must "right size," rather than releasing excess workers consider moving skilled technical employees around to other departments based on market need. Not only will you retain their knowledge base, but also the technical professionals will learn new business skills. For example, when technical products become outdated, but the employees do not, help the technical professionals find suitable positions with your organization.

68 Offer technical employees a one-month paid sabbatical for every three years of service they give your organization. This will keep employees around longer (because they know they have the extra time coming), and many will use the sabbatical for personal and/or professional development such as traveling or taking classes. They will likely return to their job more refreshed, energized and educated.

 To sway workaholics from burning out, develop a buddy system under which technical professionals monitor each other, especially when someone isolates themselves in their cube or office for days at a time. Consider hiring "get a life" counselors whose job it is to coach programmers on the joys of the real world – sunlight, bike riding, children, etc. Hewlett-Packard (in the Great Lakes area) understands work-life integration well. Employees are now asked to set annual goals not only for productivity but for leisure as well. They are expected to meet both, and if they fall short, their supervisors have to answer for it. When a staff member achieves a milestone, such as leaving at 2:00 p.m. to take a daughter ice-skating, co-workers are encouraged to support the employee with the same gusto as they would for someone landing an order for laser printers.

Creative Compensation

 Review salary and performance standards once a quarter. Keep pace with the market and keep in constant communication with your people. Also use multiple compensation vehicles like frequent bonuses, especially for "hot" skills. Also, *all* retention efforts must be reviewed and revised on a regular basis to achieve maximum success in retaining technical staff. This includes continuous assessment (and timely reassessment) of current projects in terms of short and long-term staffing projections.

"In order to keep competitive, we've made the salary structure for technical people separate. No standard 4% raises like the rest of the employees."[26]
- Sam Wheeler, Master Lock

---

26 Sam Wheeler, Vice President of Human Resources, Master Lock (Milwaukee, Wisconsin), December 1999.

71 If you really want to hold on to technical professionals, don't be too rigid in the way you approach annual merit increases. Star performers should be paid for performance against industry standards and market worth, and not restricted by internal compensation policies and procedures. Remember, when it comes down to luring one of your top performers away, your competitors will bend all the rules.

72 Compensate technical managers for their success at keeping engineers and other technical staff members on board. This gives them a specific financial incentive to effectively manage their people and improve the communication between technical supervisors and their direct reports.

73 Poll your technical managers to find out who are the most productive and/or critical employees. This group of technical employees should receive the bulk of your retention efforts, since you may not have the time or resources to lavish attention on everyone.

74 Be cautious about offering stock options as the key component of compensation – they can promote greed. You don't want to encourage people to make enough money just so they can leave. Also, a competitor can come along and offer a better stock option plan, at least on paper. When you focus all of your attention on monetary rewards, you create a culture where technical employees are only loyal to you until another competitor offers them more.

75 If you are part of a large organization, try to approximate the startup experience within your company. For example, identify projects that you consider critical to your organization's future. Then create "shares" in those projects (complete with documents resembling stock certificates) and distribute them to engineers working on relevant project teams. At the end of a fixed time period (like 3-5 years), turn those shares into cash based on the profits the projects generate. Engineers can make a lot of money, but only if they stick around.

76 If your organization is paying significantly more for technical contractors than salaried employees, then you might consider taking that extra money and spending it on your existing technical workforce. It will improve morale and minimize the salary issue when other companies try to recruit your staff.

77 Don't sit higher paid contractors next to employees. It won't take long before they talk about compensation, which will likely cause your technical employee to either ask for a raise or start looking elsewhere for a higher-paying project.

78 To help overcome the high salaries the market dictates for certain technical skills, consider designing other forms of creative compensation. For instance, introduce a loyalty scheme for technical employees. Office hours could warrant one point for each hour worked, and overtime merit two points per hour. Let them redeem their points against training, insurance, mortgage advice, etc. Or automatically contribute to every technical employee's savings account each year, no matter what the employee contributes. Make sure to communicate this benefit to the employee in monthly statements with wording like "here's another example of our investment in your future." Or consider offering an extra week of vacation in lieu of a higher raise. Although it only works out to two percent of a yearly salary, the thought of extra vacation may sound very appealing to overworked technical professionals. How about offering your technical employees pet insurance? This is especially attractive to young technical professionals who consider their pets as members of their family. Here are some examples of companies that practice creative compensation:

- FedEx Express (Memphis, Tennessee) provides free rides in the jump seat on company planes throughout the world.

- Steelcase (Grand Rapids, Michigan) has a 1,200-acre camping and recreational area for employee use. Employees have the free use of rowboats and canoes. They also offer a "take-home meal" program where employees can purchase (at cost) a five-course dinner.

- Fannie Mae (Washington, District of Columbia), while offering competitive salaries, also "sweetens the pot" by providing 10 paid hours per month for volunteer work in the community.

- Viewpoint Digital International (Salt Lake City, Utah), a Computer Associates Company, does not give anyone in their Utah-based product technology group a salary. They're still full-time employees, with benefits, but they're paid like contractors. Every project's team splits a percentage of the project based on expected revenues. Almost overnight compensation has jumped 60-70%, but productivity has almost doubled during the same time frame.

Recognition and Involvement

 Since recognition is very important to technical professionals, create a system that acknowledges superior technical contributions. Give awards to technical professionals who have accomplished a significant product milestone, went beyond the call of duty supporting a client, or who fixed a critical "bug" in the system. Make sure these awards are announced at company meetings so that the person is recognized in front of their peers. Consider establishing a "Walk of Fame" in your building with pictures of top technical performers and what they accomplished. Include the newest members of the club with an announcement in the company newsletter and corporate

Web site. By recognizing their achievements, technical professionals feel acknowledged by the organization.

80 Many technical workers are competitive. Consider sponsoring sporting challenges against other companies (running, tug-of-war, bicycling, etc.). Give the participating members company T-shirts (or some other team emblem), and provide them with metals or plaques if the team finishes first. If you can't compete against another company, consider holding competitions between different departments within your company. Not only will this promote team building, it will also get many of those individuals away from the computer monitor for some much-needed fresh air.

81 Extra-curricular activities outside of the office can also go a long way towards attracting and retaining technical professionals. For example, Documentum (Pleasanton, California) takes their technical staff to opening day of sci-fi and adventure movies on company time! They also offer lunchtime, in-house trilogies of the Star Wars series and Indiana Jones' movies, providing them with pizza/popcorn/junk food. In addition, Documentum has technical department picnics (lots of yummy ethnic food catered) at a park where the engineers can play tennis, basketball, baseball, rollerblade, etc. If your company already provides extra-curricular activities, be sure to include pictures of these events on the career section of your corporate Web site so potential candidates can see what they are missing.

Provide a system of team-based goals, not just individual goals. For example, Autodesk (San Rafael, California) corporate goals are based on team results that foster collaboration and commitment within a particular group. Each individual has a personal and team bonus based on performance. If the company does exceptionally well because of the team's work, an individual employee can potentially double his or her personal bonus. And when technical people are working well as a team, this sense of collaboration is a strong incentive to stay with the organization.

 Allow technical professionals to elect representatives from their department to participate in management or board meetings, and pay them a fee (say $50) for each meeting they attend. Elected members will feel recognized by their peers, and the level of involvement gives people a real say in the company's direction.

 Consider rearranging offices to recognize talented technical workers. Identify your most valuable performers, and allow them to have the corner and/or window offices while their managers sit in cubicles. The recognition alone is a strong motivator for the technical employee.

Fun and Engaging Environment

 A strong company culture is a competitive advantage for attracting and retaining technical talent. Creating an environment in which people want to work is crucial. If technical professionals trust the people they work with, they're more engaged and productive. A good example of strong company culture is PeopleSoft (Pleasanton, California):

- Employees boast about their "hire number" on their company T-shirt. The lower the number, the greater the pride (as well as the stock options).

- The company holds "bring your parents to work day," where parents proudly see where their children work and receive "PeopleParent" T-shirts to mark the event.

- They also have a "PeopleBaby" program. Every time an employee has a baby, he or she gets a T-shirt emblazoned with the PeopleSoft name, the baby's photo, and a baby number.

 Along with a strong company culture, your organization has to make room for the technical professional's personal values to retain their creativity, productivity, loyalty and ethical behavior. Energized, committed employees offer an organization a significant competitive edge. Technical employees will devote their loyalty and best effort if they know their company stands for values they themselves hold. How do you do this? By providing a values clarification experience for each technical professional. Help them see how their personal values relate to your organizational values, where they overlap, and how to build on the synergies.

 Providing fun activities will work wonders for attracting and retaining technical professionals. One creative activity is to put a CD jukebox and a karaoke system in the lunchroom with a complete menu of selections. Let people choose special requests and perform for their peers. Other examples of fun activities include:

- A slide from the 2nd floor to the 1st floor kitchen has been installed at Berkeley Systems (Berkeley, California). Employees can choose to use it rather than the stairs.

- Golf driving range nets have been installed at Kingston Technology (Fountain Valley, California), so that technical workers can work on their golf game without leaving the facility.

- A telecommunications company in Cleveland, Ohio has foosball in the company's recreation room, and technical employees compete each year for a pair of World Series tickets.

- 3dfx (San Jose, California) has established "dog days" for technical employees to bring their dogs to work if they choose. They see all shapes, sizes and colors, and everyone has a great time with man's best friend. They also translate this benefit to their recruiting efforts by placing large posters throughout company buildings that display a picture of a playful dog and the caption "This isn't the only friend you can bring to work."

- Genentech (South San Francisco, California) sends teams of technical employees to cooking class to promote bonding.

- Every Thursday at 3:00, Gymboree Corp. (Burlingame, California) rings a bell and declares a 30-minute recess. Some people take walks while others play on the grounds. 4-square is very popular, with different departments competing against each other.

 If your technical staff is frequently pulling all-nighters, put up a tent in a dark corner of the office, and include a pillow, mat, sleeping bag, alarm clock, eyeshade and portable tape machine with relaxation music. Call it the "relaxation area" where technical people can go for a few minutes and unwind. This is one way to help them deal with their stress. Or, if you want to splurge, consider an example by 3dfx (San Jose, California). They found that their engineers prefer to work late into the night. Therefore, they established the "Voodoo Lounge" that has large leather recliners and sofas. It is also equipped with a fancy stereo system and big screen for their technical staff's enjoyment, and is fully stocked with free beer, soda, teas, bottled water and snacks.

 Humor works wonders to relieve stress associated with tight product deadlines, and there are several ways to add humor to the technical environment. For example, you could create a humor board where people can pin up their favorite joke, comic strip, card, etc. Or you could have an ugly clothing contest, which brings out the creative side in people. How about an event like a beach party during which your lunchroom is equipped with a karaoke machine and Super Soaker™ water guns? Another favorite is the whimsical game Twister™ that lets participants square off in a battle to see who has the best balance while stretching in funny positions. Tap into the kid inside of you for other fun ideas.

 Subsidize cafeterias so that workers don't have to worry about meals, and allow family members to come in and eat dinner together with employees free of charge when they have to work late. This convenience will pay off with greater productivity and loyalty from the technical professional.

ENLIGHTENED
ORGANIZATION EXAMPLES

The following are two examples of enlightened organizations that truly understand the importance of investing resources to keep their best and brightest people.

Cisco Systems, Inc.

Cisco Systems, Inc. (San Jose, California) is the worldwide leader in networking equipment for the Internet, and much of the credit for their success goes to their ability to attract and retain top talented technical people. Cisco's culture and work environment is so stimulating that headhunters acknowledge that unless the job opportunity is with a start-up, it is nearly impossible to pry someone away from Cisco (www.Cisco.com).

Some of the elements that have lead to a strong company culture include the following:

- A "Fast Start" orientation program, designed to achieve the fastest time to productivity for new hires in the industry. Cisco considers the new employee's first day the most important 8 hours of his or her career. To optimize new employee productivity:

- Employees arrive on their first day to find their office completely configured with the necessary computer and telecommunications equipment so that they can be productive on day one.

- Using the Cisco employee Intranet, new hires can visit an online "dashboard" that provides them with all the information they need to set up benefits online, and takes them through a list of tasks to complete within their first few days and weeks on the job.

- New employees are assigned a mentor to help them with their transition into the new job.

- Cisco believes in spending money in two areas: customers and employees. Stock options are distributed generously, with a full 60% of all Cisco stock options in the hands of individual contributors not at the management level.

- Individual contributions are widely recognized and celebrated around the company. Employees can reward another employee's exceptional performance with an on-the-spot bonus that ranges from cash to tickets to a special event.

- Their Web site for attracting talent (see Virtual Strategies) is so user-friendly that it gives them a competitive advantage for attracting the passive job seeker.

- Cisco's CEO, John Chambers, addresses all employees during a company meeting every quarter. The meeting serves as a company report card and a way for any employee to ask questions in an open forum fashion. Employees not located in the San Jose facility are also included either by Webcasting, satellite TV or by recording the meeting and reviewing it later. This is one example of their open communication management style.

SAS Institute

SAS Institute (Cary, North Carolina), the world's largest privately held software company and the leading provider of data warehousing and decision-support software, is a good example of establishing an environment conducive to attracting and retaining technical professionals.

As you read through the list of benefits SAS Institute provides, you may wonder why they do all this. It's because senior management believes that when employees are treated well, they treat the company and the customer well. With an average SAS salary of $50,000 and a replacement cost of at least 1.5 times salary, the company says it saves $67.5 million per year compared to their competition, which means they have $12,500 extra per employee to spend on benefits. This strategy has also paid off in a low employee turnover rate of 4%, as well as a customer retention rate of 98%.[27]

SAS Institute (www.SAS.com) believes in long-term relationships with employees and customers. A low turnover rate enables the company to develop product faster because development teams stay together and working relationships are preserved. Also, customers develop relationships with the company through devel-

---

27 Les Hamashima, Manager, Public Affairs Department, Corporate Communications Division, SAS Institute (Cary, North Carolina), January 2000.

oping long-term relationships with employees. This is one of the reasons SAS Institute wins technical support awards.

SAS employees enjoy the following:

- Child care for $200 per month
- The largest on-site daycare operation in the state (and one off-site facility)
- A 35-hour full-time work week
- Live piano music in the cafeteria (which also offers gourmet food!)
- Unlimited soft drinks
- Hundreds of pounds of M&M's are distributed to every floor of every building
- One week paid vacation between Christmas and New Years
- A 36,000 square foot on-site gym that includes pool tables, ping-pong, basketball, aerobics, cardio machines and weights (dirty workout clothes are laundered overnight for free!)
- On-site health clinic staffed with 6 nurses and 2 physicians
- Zero cost to employees for health insurance
- Casual dress every day (except in client interactions)
- Elder care advice and referrals
- On-site massages several times per week
- All family benefits extended to domestic partners, regardless of sexual orientation
- Unlimited sick days.

RECOMMENDED RESOURCES

To find out more about the subject of attracting and retaining technical talent, we recommend reviewing the following books, articles and Web sites:

Books

➤ *First, Break All The Rules* by Marcus Buckingham and Curt Coffman, Simon & Schuster, 1999.

This book is the product of two mammoth research studies undertaken by the Gallup Organization over the last 25 years, asking employees from a broad range of companies and industries questions about the most important needs demanded by the most productive employees.

➤ *The Human Equation* by Jeffrey Pfeffer, Harvard Business School Press, 1998.

The author explains why even smart organizations fall into harmful patterns when managing people, and offers specific, actionable steps that can be taken to achieve improved performance.

➤ *Working With Emotional Intelligence* by Daniel Goleman, Bantam Books, 1998.

The author reveals the skills that distinguish the star performers in every field. From entry-level jobs to top executive positions, Goleman explains that the single most important factor is not IQ, advanced degrees or technical expertise. Rather, it is emotional intelligence.

➤ *1001 Ways To Reward Employees* by Bob Nelson, Workman Publishing Company, 1994.

A classic handbook for generating ideas to keep your employees motivated.

Articles

➤ "Job Boom Brings Insecurity: Report Says Positions Are Short-Term, Low Pay" by Ramon G. McLeod, *San Francisco Chronicle*, May 26, 1999.

A summary of the results of a joint study by Working Partnerships USA and the Economic Policy Institute that found that people are being forced to change jobs more rapidly.

➤ "How To Manage Geeks" by Russ Mitchell, *Fast Company Magazine*, June 1999.

A 9-point tutorial by the CEO of Novell.

➤ "Sanity Inc" by Charles Fishman, *Fast Company Magazine*, January 1999.

An overview of the employee recruiting and retention program of SAS Institute.

➤ "Organizational Development Forum: Managing In a Virtual Environment" by Rick Barton and Mary Ann Bopp, *Issues & Trends Report*, American Society for Training & Development, 1998-1999.

A good overview of the organizational issues involved with offering telecommuting and other virtual management strategies.

➤ "How're You Going To Keep 'Em Down On the Firm" by Christopher Caggiano, *Inc. Magazine*, January 1, 1998.

Talks about how one firm kept its most important employees.

➤ "How Recruiters Woo High Demand Candidates" by Robert J. Grossman, *HR Magazine*, December 1998.

An overview on the importance of recruiting and retention to an organization's success.

➤ "Keep Them!" by Charlene Marmer Solomon, *Workforce Magazine*, August, 1997.

Includes a formula for determining the return on investment (ROI) for your retention efforts.

➤ "Desperate Times, Creative Measures" by Peter Fabris, *CIO Magazine*, January 1, 1998.

Additional strategies to recruit and retain technical professionals.

➤ "The Soft Touch" by Matt Vilano, *CIO Magazine*, March 15, 1999.

What to do if you lack financial resources to offer big salaries and signing bonuses.

➤ "Hot Skills and Cold Cash" by Debby Young, *CIO Magazine*, March 15, 1999.

A good overview of what types of cash bonuses to offer technical professionals.

➤ "How To Bridge the Loyalty Gap" by Dr. Edward Wakin, *Beyond Computing*, June 1998.

What employees want, besides salary, from their employers.

➤ "Recruiting and Retaining Technical Talent: Local Solutions To the Global Crisis" by Peter Alexander, *Technology Business*, November/December 1998.

A quick summary of some of the tactics and issues discussed in this book.

➤ "A New Alliance: Recruiting and Retaining Good Technical Employees Is a Job Requiring a Partnership of HR and IT" by Jodi Spiegel Arthur, *Human Resource Executive Magazine*, October 19, 1998.

An excellent overview of strategies used by several companies in various industries.

➤ "Name That Salary: With Their Skills In Demand, IT Workers Can Write Their Own Ticket" by Michael A. Verespej, *Industry Week*, February 15, 1999.

The article talks about the role compensation has with technical professionals.

➤ "Bosses From Heaven and Hell" by Bronwyn Fryer, *ComputerWorld*, August 9, 1999.

An article that talks about the perception IT workers have about their supervisors.

➤ "The Care and Feeding of Technical Professionals" by Tim Walsh, *Employment Management Today* (Society of HR Management), Winter, 1998.

A good overview of the issues involved with retaining technical professionals.

Web Sites

➤ www.CorporateAlumni.com helps companies keep in touch with former employees, and former employees to keep touch with one another. This site can help keep your organization in touch with bright people who have moved on to other challenges.

➤ www.RecruitersNetwork.com and www.InterBizNet.com offer free monthly e-mail newsletters with fairly useful information about Internet recruiting trends, job boards and related topics.

➤ www.MyBoss.com is a place for workers around the world to post anecdotes about their bosses, many of which are not very positive. This site will give you an idea of the power of the Internet to spread positive or negative word-of-mouth information regarding an organization or manager.

➤ www.Vault.com has 1,000 chat rooms where employees of companies talk about the management style and culture of their organization. Technical professionals can find out the real story about a company or manager. Log in and see if your organization is being talked about.

➤ www.WorkforceOnline.com is the electronic version of *Workforce Magazine*, and it has an easy-to-use search function to find ideas for attracting and retaining employees.

➤ www.WorkingWounded.com is a place for people to come together to laugh and learn how to survive today's turbulent workplace.

SUGGESTION BOX

If you have a tip or suggestion for future volumes of this book, we would love to hear from you. If we include your suggestion, we will give you free press and publicity by citing the source of the tip. All you have to do is e-mail us at info@bwinc.com from anywhere in the world. Or fill out the form below, and then fax or snail-mail it to us at the following address:

Blessing/White, Inc.
Attn: *Money Isn't All That Matters* Handbook
c/o Marketing Department
23 Orchard Rd., Ste. 2, Skillman, NJ 08558-2609, USA
Phone: +1-908-904-1000 • Fax: +1-908-904-1774
Web: www.BlessingWhite.com

Name: _____

Company: _____ Title: _____

E-mail Address: _____

Phone Number: _____

Suggestion/Tip: _____

_____

I authorize Blessing/White to include my suggestion in future editions of *Money Isn't All That Matters*. (Please note: we require your signature to include your suggestion. Thanks!)

Signed: _____

ABOUT THE AUTHORS

DAN TREADWELL

Dan Treadwell, a Senior Consultant for Blessing/White, has spent the last 8 years of his career focusing on the technology sector. A former technical leader himself, Dan devotes the majority of his professional time helping technical leaders and professionals improve their management and communication skills and approaches.

A graduate of The College of Wooster in Wooster, Ohio, Dan has managed technical and non-technical workers in capacities as diverse as marketing, accounting & finance, consulting and advertising in industries such as performing arts, banking, advertising, sales promotion, and training.

As a Consultant, Dan has traveled extensively throughout the continental US, Canada and the UK designing training for a wide range of technical audiences in industries such as computer hardware, computer software, finance, telecommunications, aerospace, manufacturing and the federal government.

PETER ALEXANDER

 Peter Alexander studies market trends and developments regarding the "people side" of high technology. Personal experience with a software developer who went bankrupt because of poor communication convinced Peter that there is more to being a successful business than "getting product out the door."

He has been featured in several national publications, including *Technology Business, InformationWeek, Marketing News* and *Human Resource Executive*. Peter also serves as an Instructor for the University of California, Berkeley Extension program. He holds B.S. and M.B.A. degrees in Business Management from the California State University system.

In his career, Peter has worked in management roles within various industries including Internet strategy, software development, medical equipment and transportation logistics.